Multiple Fatality
Single-Family Dwelling Fire
St. Cloud, Florida

Investigated by: John Lee Cook, Jr.

This is Report 142 of the Major Fires Investigation Project conducted by Varley-Campbell and Associates, Inc./TriData Corporation under contract EME-97-CO-0506 to the United States Fire Administration, Federal Emergency Management Agency.

Department of Homeland Security
United States Fire Administration
National Fire Data Center

U.S. Fire Administration Fire Investigations Program

The U.S. Fire Administration develops reports on selected major fires throughout the country. The fires usually involve multiple deaths or a large loss of property. But the primary criterion for deciding to do a report is whether it will result in significant "lessons learned." In some cases these lessons bring to light new knowledge about fire--the effect of building construction or contents, human behavior in fire, etc. In other cases, the lessons are not new but are serious enough to highlight once again, with yet another fire tragedy report. In some cases, special reports are developed to discuss events, drills, or new technologies which are of interest to the fire service.

The reports are sent to fire magazines and are distributed at National and Regional fire meetings. The International Association of Fire Chiefs assists the USFA in disseminating the findings throughout the fire service. On a continuing basis the reports are available on request from the USFA; announcements of their availability are published widely in fire journals and newsletters.

This body of work provides detailed information on the nature of the fire problem for policymakers who must decide on allocations of resources between fire and other pressing problems, and within the fire service to improve codes and code enforcement, training, public fire education, building technology, and other related areas.

The Fire Administration, which has no regulatory authority, sends an experienced fire investigator into a community after a major incident only after having conferred with the local fire authorities to insure that the assistance and presence of the USFA would be supportive and would in no way interfere with any review of the incident they are themselves conducting. The intent is not to arrive during the event or even immediately after, but rather after the dust settles, so that a complete and objective review of all the important aspects of the incident can be made. Local authorities review the USFA's report while it is in draft. The USFA investigator or team is available to local authorities should they wish to request technical assistance for their own investigation.

This report and its recommendations were developed by USFA staff and by Varley-Campbell & Associates, Inc. Miami and Chicago, its staff and consultants, who are under contract to assist the Fire Administration in carrying out the Fire Reports Program.

The Federal Emergency Management Agency, United States Fire Administration gratefully acknowledges the cooperation of the Fire Chief and members of the City of St. Cloud Fire Rescue Department. Everyone who assisted in the preparation of this report was generous with his or her time, expertise, and counsel.

For additional copies of this report write to the U.S. Fire Administration, 16825 South Seton Avenue, Emmitsburg, Maryland 21727. The report and the photographs in color are available on the Administration's Web site at http://www.usfa.dhs.gov/

U.S. Fire Administration
Mission Statement

As an entity of the Department of Homeland Security, the mission of the USFA is to reduce life and economic losses due to fire and related emergencies, through leadership, advocacy, coordination, and support. We serve the Nation independently, in coordination with other Federal agencies, and in partnership with fire protection and emergency service communities. With a commitment to excellence, we provide public education, training, technology, and data initiatives.

TABLE OF CONTENTS

OVERVIEW 1
SUMMARY OF KEY ISSUES 2
THE ST. CLOUD FIRE RESCUE DEPARTMENT 2
BUILDING CONSTRUCTION AND OCCUPANCY 3
THE FIRE 3
THE INVESTIGATION 6
LESSONS LEARNED (RELEARNED) 7
APPENDIX A: Maps and Diagrams 9
APPENDIX B: List of Photos 15

Multiple Fatality Single-Family Dwelling Fire St. Cloud, Florida

Local Contacts: Charlie Lewis, Fire Chief
City of St. Cloud Fire Rescue Department
4700 Neptune Road
St. Cloud, Florida 34769
407-891-6781

Bill Johnston, Assistant Chief
407-891-6782

Ron Hood, Lieutenant
Incident Commander

Jim Grenus, Acting Officer
First Arriving Engine Company

OVERVIEW

On June 21, 2001 a fire in a single-family detached dwelling resulted in the death of six occupants, five of whom were children ranging in age from twenty-two months to seven years old. The fire was reported at approximately 2:00 a.m. and everyone was reported to have been asleep at the time that the fire was discovered. According to the sole surviving resident, there was no operational smoke detector in the home. As a result, the fire reached sufficient intensity as to make the rescue of the victims impossible. By the time firefighters arrived, the structure was heavily involved and fire had broken through the doors and windows and was threatening nearby exposures.

Firefighters quickly extinguished the blaze and were left with the grim task of searching for and recovering the bodies of the six victims: a twenty-four year old female, her twenty-two month old baby girl; the five-year-old son and seven-year-old daughter of her housemate; and the six-year-old son and seven-year-old daughter of her housemate's sister.

Investigators determined that a battery charger had been left plugged into an electrical outlet on the enclosed front porch of the dwelling and had caused the fire. The battery charger was not in use at the time of the fire, but had been covered by clothes and apparently overheated, igniting the blaze.

Fire fatalities are rare in St. Cloud and the number and age of the victims deeply impacted both the firefighters and the community as a whole. A number of fundraisers were held to assist the families of the victims and a local funeral home donated the funeral services. The demand for smoke detectors also dramatically increased immediately following the fire.

SUMMARY OF KEY ISSUES

Issues	Comments
Absence of a Functioning Smoke Detector in the Residence	While a smoke detector was reported to have been in place when the occupants rented the house, the sole survivor indicated that it was not operable at the time of the fire. The absence of an operational smoke detector continues to be a significant factor in residential fire fatalities. The National Fire Protection Association (NFPA) reports that 3,445 fire deaths occurred in the home in 2000, eighteen percent more than the year before. Residential fires accounted for 75.1 percent of all structure fires and 85.2 percent of all deaths (72.2 percent occurred in single-family dwellings).
	According to the NFPA, the term catastrophic fire refers to fires that kill five or more people in a residential property or three or more in a nonresidential or nonstructural property. In 2000, residential fires were responsible for eight of the thirty-four catastrophic fires (53%). Fifteen occurred in single-family dwellings and three occurred in apartment buildings. Combined, the fires killed ninety-nine people. Twenty-five of the victims were children under the age of six. Ten of the eighteen homes did not have a smoke alarm and one had an alarm with no battery. Sixteen of the fires occurred between 11:00 p.m. and 07:00 a.m..
Age of the Victims	Five of the victims were children, the eldest of which was seven. Three of the victims were five or younger. Although children five and younger make up only nine percent of the total population, they comprise nineteen percent of all fire fatalities according to the NFPA. The United States Fire Administration has specifically targeted children fourteen and younger in a special effort to reduce the number of fatalities and injuries in this age group.
Community Impact	The number and the age of the victims made this a particularly difficult event for not only the firefighters, but for the community as well. Following the fire, a number of fundraisers were held to assist the families of the victims and a local funeral home donated funerals for each of the victims. There also was an immediate, but temporary surge in the demand for smoke detectors.
Complacency	The St. Cloud Fire Department had given away smoke detectors for years prior to this incident. The demand had been relatively low to the point that the Department was experiencing problems with batteries going bad before the detectors could be given away.
Time of Day	The fire was reported at 2:07 a.m.. All of the occupants were asleep at the time the fire was discovered and the absence of an operational smoke detector allowed the fire to burn undetected until it grew in size and intensity to prevent the escape of the occupants in the upstairs bedrooms.

THE ST. CLOUD FIRE RESCUE DEPARTMENT

The City of St. Cloud is located in the central part of the Florida Peninsula in Osceola County. St. Cloud was founded in 1909 as a land development to improve properties south of the City of Orlando. Billed for its health, climate, and productiveness of the soil, the Town of St. Cloud was incorporated in 1911 as a soldiers' colony for veterans of the Civil War. St. Cloud is a diverse community of approximately 20,000 residents and covers a growing 8.5 square mile service area.

Fire and emergency medical services are provided by the City's Fire Rescue Department, which is an all career service with an annual budget of approximately $3 million. Firefighters are deployed in two stations and work a rotating 24-on/48-off shift. Each of the three platoons staff two engine companies and two ALS rescue units (ambulances). Normally, minimum staffing is ten, which includes an operations shift commander. The department responds to approximately 3,000 incidents each year with EMS calls constituting 78% of the total call volume. The normal alarm assignment for a structure fire is the entire on-duty shift.

BUILDING CONSTRUCTION AND OCCUPANCY

The house was located at 1409 Maryland Avenue, between 14th and 15th Streets and was a two-story single-family detached dwelling with a detached one-car garage. Built in the 1920's, the wood-frame structure had a metal roof. Based upon the appearance of the home (see photo no. 4, page 18) it is believed that the upstairs portion of the house had been enlarged at some point. Florida has a very mild climate and the washer, dryer, and hot water heater were all located on the back porch, which was open on three sides.

The house also had a front porch, which had been enclosed at some point. Total square footage was approximately 1,500 square feet including the enclosed front porch, but excluding the open back porch (see diagram no. 1, page 12). There was a bedroom, bathroom, living room, and kitchen on the first floor and two bedrooms and a bathroom on the second floor. A single, interior straight stairway provided access to the second floor.

THE FIRE

A 9-1-1 call was received by the St. Cloud Emergency Communications Center at 02:07 a.m. from a male occupant of 1409 Maryland Avenue who reported that his house was on fire and that there was the possibility that a number of the occupants were trapped in the burning structure. Telecommunicators dispatched a structure alarm assignment consisting of two engines, two rescues, and the operations shift commander. A total of ten firefighters responded on the initial alarm.

St. Cloud Station One is located nearby (see map 2, page 11) at 915 Massachusetts Avenue and arrived within four minutes of being dispatched. Commanded by an acting officer; the firefighters could see heavy flames two blocks from their location and radioed the operations shift commander. The commander requested that additional companies be dispatched due to the magnitude of fire and the travel distance of the second due engine company. The Osceola County Fire Department, upon receipt of a request for assistance, dispatched a career engine company, a rescue (ambulance), an attack (squad), a battalion chief, and a volunteer engine company. The volunteer engine company was later cancelled while still enroute, but the other companies continued on to the fire location and assisted with extinguishments and recovery of the victims.

There were eight people who resided at 1409 Maryland Avenue and seven were at home when the fire occurred. Six of them died. A woman, aged twenty-four, lived in the house with her twenty-two month old baby girl and both perished. She was babysitting the other children who were in the home when the fire occurred and shared the home with another woman, aged twenty-six. Her housemate lived in the home with her five-year old son and seven-year-old daughter. The twenty-six year old female was at work when the fire occurred. Both of her children died.

The twenty-six year old female's younger sister worked at the same nightclub and had left her six-year-old son and seven-year-old daughter in the care of the twenty-four year old female while they both were at work. Her children perished as well. The brother of the two club employees, aged twenty-three, also lived in the house. He too was at home at the time of the fire and is the only person to have survived.

St. Cloud Engine 101, with a crew of three firefighters, was the first company on the scene and took up position on the west side of the house on Maryland Avenue, almost directly in front of the burning structure. The Department's standard operating procedure is to use the water from the tank (750 gallons) on the first apparatus to make the initial attack and to have the second company lay the supply line, if needed.

While Engine 101's crew was pulling an attack line, the officer began to make a 360-degree size-up, but encountered a downed power line on the south side of the structure. Unable to complete his walk around, the officer ordered that the 1-3/4-inch attack line be used to protect the exposure on the north side of the fire because the siding on the exposed single-family dwelling was beginning to melt.

Upon arrival, the firefighters had been told by the occupant of the exposure on the north side of the house that there were five kids inside and that an adult male had escaped with minor cuts and smoke inhalation. Fire was visible in every window of the house and there was heavy fire on the enclosed front porch, which led firefighters to believe that no one could possibly be alive in the house. The volume and intensity of the fire made the structure untenable. Since a rescue was out of the question, their efforts were directed at extinguishment and the recovery of the bodies of the victims.

A second 1-3/4-inch attack line was deployed to protect the single-family dwelling on the south side of the fire. After the exposures were adequately protected, firefighters began to attack the fire in the downstairs portion of the house and quickly began to achieve knockdown. Meanwhile, three police officers assisted in the extinguishments effort by dragging a 5-inch supply line approximately 300 feet to the hydrant (color coded green) that was located on the northeast corner of 14th Street and Maryland Avenue.

When the second St. Cloud engine arrived; that crew pulled a 2-1/2-inch pre-connect from Engine 101 and made an exterior attack on the second floor. The operations shift commander was the next to arrive on the scene and he established command using the department's ICS procedure. After achieving approximately an 80% knock-down from the exterior, firefighters donned their SCBA's, entered the structure and conducted a primary search of the first floor. Not finding anyone, they advanced a 1-3/4-inch line up the interior stairwell not realizing that there was a victim at the base of the stairs. All exterior attack lines were shut down as the interior attack began.

The firefighters knocked down the fire on the second floor and discovered what they initially thought was two victims. Upon entering the second bedroom, firefighters found two more bodies by the rear window (see diagram 1 for the location of the victims, page 12). At this point, the IC requested that the department's Chaplin, the Medical Examiner from Orange County; the State Fire Marshal; and the local fire marshal be dispatched to the incident.

Firefighters began to exhaust the contents of their first bottle of air and backed down the stairs at which point they were advised by the officer of the Engine 201 that there was an additional victim at the base of the stairwell, bringing the total victims to five. The body had been difficult to recognize due to severity of burns.

The sixth victim, a baby girl, was later found under the body of her mother who had been found lying face down. Firefighters have postulated that she had tried to cover the baby with her own body in a vain attempt to save her. It is believed that the mother could have easily escaped since she was sleeping in the downstairs bedroom, but had taken her baby upstairs to try and rescue the four kids that were sleeping upstairs.

The weather at the time of the fire was not an issue. Skies were clear with a light breeze and the temperature was in the low 70's.

After the fire had been extinguished, firefighters learned that the adult male occupant of the house had been asleep on a couch in the living room. He woke up, discovered the fire, and called 9-1-1. According to him, there was no operating smoke detector to alert the occupants that the house was on fire. He allegedly went upstairs, but was driven back by heavy smoke. Before exiting the second floor, the man stated that he broke a window that he could not open. He then went back downstairs and exited via the rear of the structure, leaving the door open. Then, he climbed up to roof of the back porch, broke out the window, and called to the children. When he left the house, he had picked up a portable phone, which he used to call 9-1-1 from the roof of the back porch. Records indicate that a total of three 9-1-1 calls were made from the burning residence.

The dispatcher reported that she could hear the children screaming in background. Allegedly, the fire had flashed over and knocked the man over backwards. Suffering from smoke inhalation, cuts, and burns, the man then exited the roof of the porch.

An attendant at a gas station located two blocks away told fire officials that he was taking out the trash when he saw the fire and called 9-1-1. He stated that he then ran to the location with a portable fire extinguisher and attempted to extinguish the blaze. Several neighbors also used garden hoses in an attempt to bring the fire under control. After the attempt at extinguishing the fire failed, the attendant got the male occupant who had escaped away from fire.

Fire officials were told that the adults slept downstairs and that the kids all slept upstairs. The mothers of four of the children who died in the fire worked at a local nightclub and were at work when the fire occurred. When they were informed of the fire, they went to the scene and had to be restrained. One of them became so distraught that she had to be sedated and transported to a hospital. Other family members also began arriving and the scene quickly became very emotional. The police had to tape off the area for crowd control. A large number of the people in the neighborhood also began to congregate at the scene.

The St. Cloud Fire Chief was not in town when the fire occurred. Therefore, the Assistant Chief responded to the fire and functioned as the Public Information Officer, while the operations shift commander continued in his role as Command. The media became an issue even though they responded late in the incident. Operations had been conducted on a county radio channel because of mutual aid. The channel is not normally monitored by the media. Once the media learned of the incident, they began to congregate at the scene. A number of satellite television trucks showed up and the major television networks and the print media began to call and to make inquiries.

After extinguishment, firefighters covered the remains of the victims and secured the scene until the arrival of the Medical Examiner. They then assisted with the removal of the bodies, which were placed in body bags. Firefighters held up sheets to screen the victims from the vision of the many onlookers and the members of the media that had gathered at the scene. The bodies of the victims were all transported to the morgue in Orlando.

At 7:30 a.m. the next morning, the on-coming shift relieved the crews at the scene of the incident. By 10:00 a.m., all but one of the companies was released to return to service. A single engine company remained at the incident site in order to secure the scene.

Table One – Chronology of Events

Thursday, June 21, 2001

Time	Event
02:07	9-1-1 call reporting fire and possible entrapment at 1409 Maryland Avenue
02:11	Engine 101 and Rescue 101 on location, reports working fire and requests mutual aid
02:14	Engine 201 and Rescue 201 on location

Friday June 22, 2001

Time	Event
05:00/06:00	Assisted ME in removal of victims
07:30	On-coming shift relieved fire crews at scene
10:00	One engine held, others released
10:50	Last unit, Engine 101, cleared the scene
15:30	Held press conference

Because of the age and condition of structure, which was rendered uninhabitable as a result of the fire, the house was destroyed following conclusion of the investigation. A demolition permit was issued on July 2, 2001.

While there were no problems evident at scene, a number of the firefighters had young children of their own. To mitigate the possibility of a delayed reaction to the stress of the incident, the department held a CISD debriefing, which included the dispatchers and the members of the fire and police departments as well as the sheriff's department. Chaplains and firefighters with CISD training conducted the debriefing.

The incident was very difficult for many of the emergency responders to handle and fire officials stated that they will always wonder what could they have done differently that would have prevented this, or possibly saved someone? As a matter of routine, the fire department had made smoke detectors available to anyone that wanted one, but the low demand had resulted in a problem with the batteries expiring before the detectors could be given away.

Immediately following the incident, there was a very positive community response (see photos of teddy bears). A number of fund-raisers were held for the victims and their families and a local funeral home donated funeral services. There also was very positive coverage of the incident by the local media and the demand for smoke detectors increased dramatically, for a brief period of time following the incident.

THE INVESTIGATION

The fire was investigated by the State Fire Marshal's Office and the St. Cloud Fire and Police Departments. After ruling out all other causes, including possible criminal activity, investigators concluded that the fire was caused by a battery charger (see photos, page 40) that had been left plugged in to an electrical outlet on the front porch. The battery charger had not connected to anything at the time of ignition, but had been covered by clothes, which apparently overheated and then ignited.

In Florida, an investigator from the State Fire Marshal's Office routinely investigates any fire involving a fatality. Local fire marshals do not have police powers. Therefore, the St. Cloud Police Department assisted in the investigation. Due to the sensitive nature of the loss, police provided both a crime technician and an investigator to collect evidence and to assist in the determination of cause and origin.

Officials also used an accelerant dog and took a number of samples from the burned structure. The investigation was prolonged for several days as investigators brought in an electrical expert to assist them in making their determination of the cause and origin of the fire.

LESSONS LEARNED (RELEARNED)

1. **The absence of a functioning smoke detector within the residence contributed to the severity of the incident.**

    Hearsay evidence suggested that the smoke detector had been disconnected three days prior due to the fire due to a malfunction. The owner of the home was able to document that a functioning smoke detector had been present when the house was occupied. Evidence continues to suggest, that an operable smoke detector is one of the best tools in the prevention of death in a fire in a residence and that the absence of an operable smoke detector continues to be a factor in the numerous deaths that do occur.

    The local fire department had a very aggressive public education program in effect prior to the fire that included providing free smoke detectors and replacement batteries. Prior to the fire, batteries would often expire before they could be given away. After the fire, the average daily demand increased to twelve smoke detectors and six batteries per day.

2. **The time of day worked to the disadvantage of the emergency responders.**

    The fire was reported at approximately 2:00 a.m. when the majority of occupants of a residence would normally be asleep. The absence of an operational smoke detector contributed to both a delay in the discovery of the fire and allowed the fire to become large enough to prevent all but one of the occupants from escaping. Data collected by both the USFA and the NFPA continue to demonstrate that time of day and the absence of an operational smoke detector is a deadly combination, which contributes to a significant number of fatalities in residences each year in the United States.

3. **Socioeconomic factors continue to be a major factor involving fire fatalities in residential occupancies.**

    Fires continue to kill the very young, the very old, and those in the lower end of the socio-economic spectrum in disproportionate number when compared to the general population. Five of the six victims were seven or younger. The home was located in a modest, working class neighborhood, once again meeting the profile that all too commonly present in catastrophic events such as these.

    People at socio-economic disadvantage are more likely to share living quarters designed for fewer occupants, thereby increasing the rescue and response burden to the fire department.

APPENDIX A

Maps and Diagrams

Map One: Location of the fire: 1409 Maryland Avenue

Map Two: Location of St. Cloud Fire Station One proximity to the fire

Diagram One: Sketch of the Floor Plan

Diagram Two: Relationship of the victims

Map One: Location of the fire: 1409 Maryland Avenue

USFA-TR-142/June 2001   11

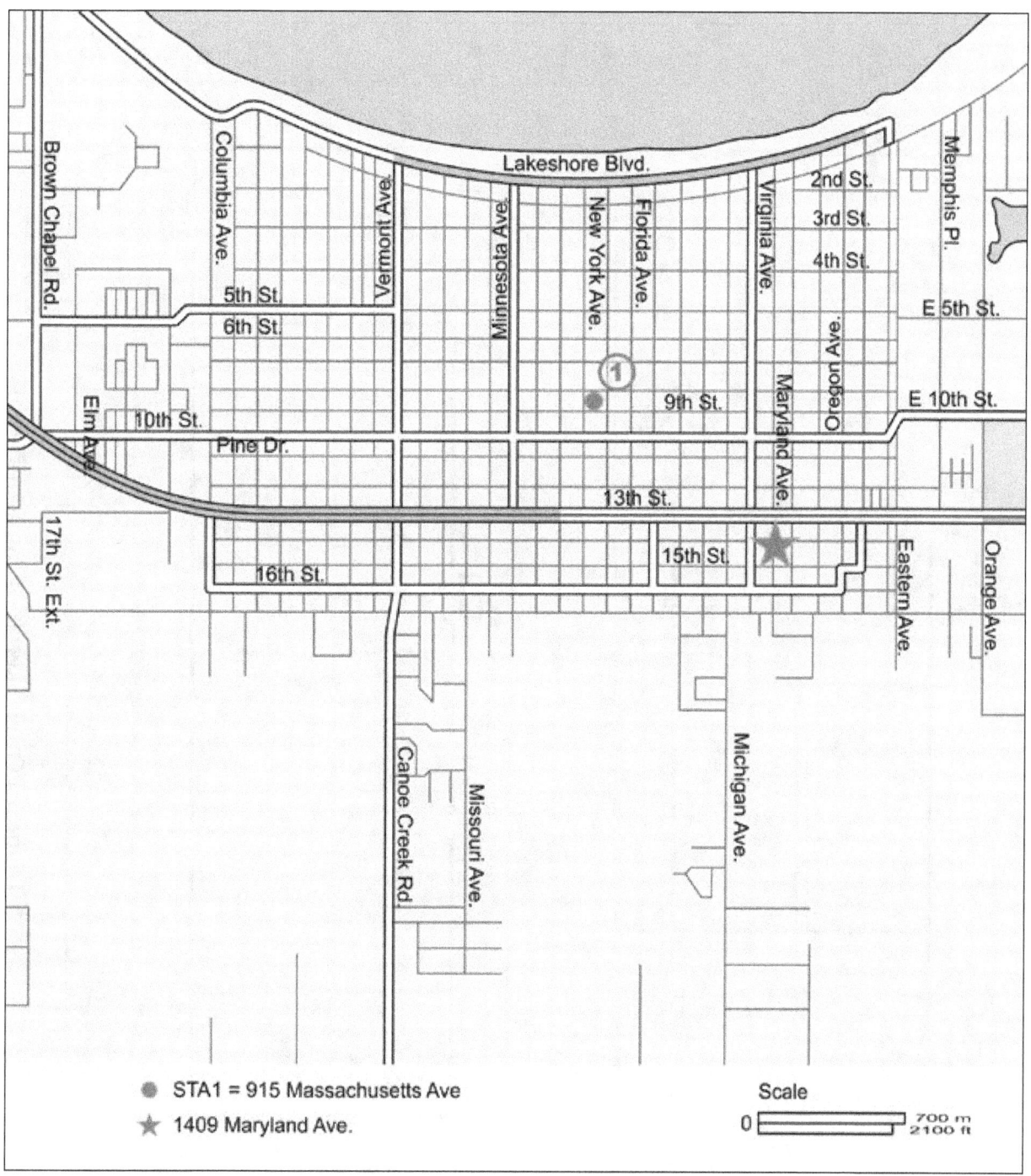

Map Two: Location of St. Cloud Fire Station One proximity to the fire

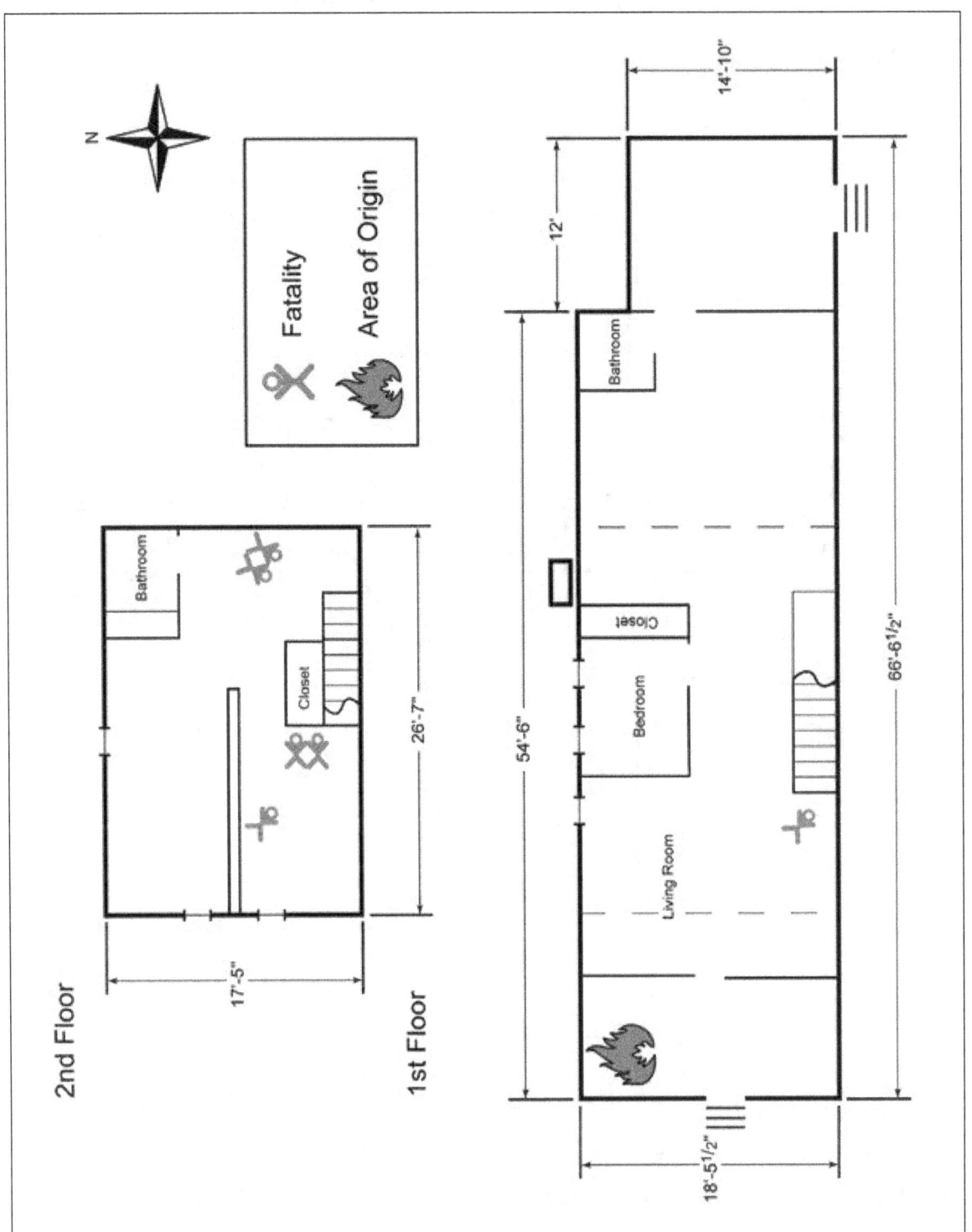

Diagram One: Sketch of the Floor Plan

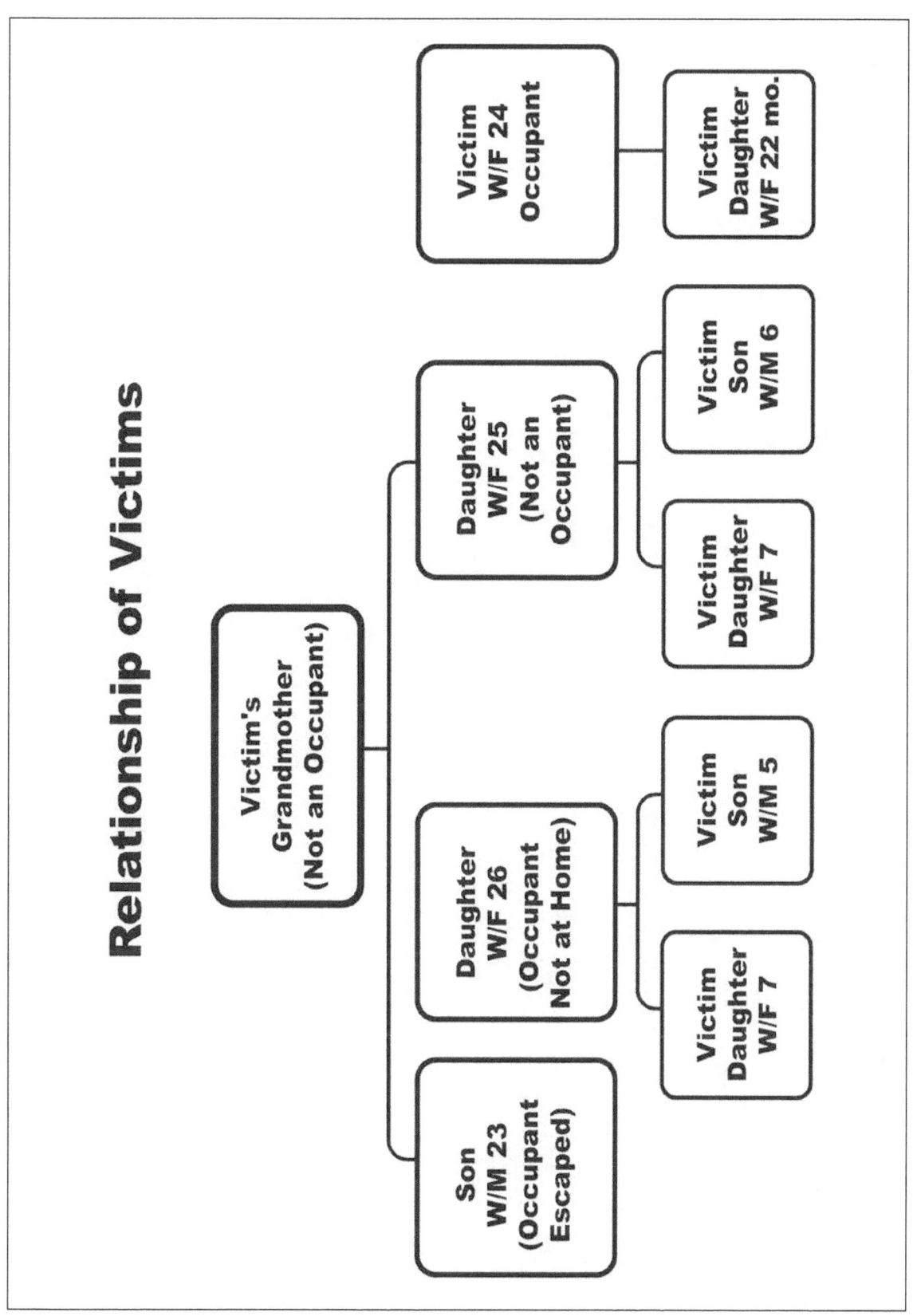

Diagram Two: Relationship of the victims

APPENDIX B
List of Photos

Unless otherwise indicated, all photos were furnished by the St. Cloud Fire Department.

Photo	Description
#1	Night view of northwest corner (front) of house
#2	Night view of front entrance on west side of house; note firefighters and hose lines on front porch
#3	Daylight view of front of house; investigators at work
#4	View of north side of house; note multiple roof lines indicating possible enlargement of second floor
#5	Day time view of south side of structure; open back porch is located behind lattice; note electric meter in center of wall to the left of the second window
#6	Day time view of rear of house and open back porch; note: this is where male occupant climbed up and broke out rear window and called for children; two bodies located just inside the rear window on second floor
#7	Exterior view of enclosed front porch
#8	Exterior view of second floor window of rear bedroom; note charring pattern
#9	North view of exterior; note ground ladder extended to second floor
#10	Another view of north wall of house; note ladder and char patterns above all three windows
#11	Night view of southern wall; again pattern of exterior side indicative of former enlargement of second floor
#12	Northern exposure; note melting of exterior vinyl siding
#13	Northern exposure; note melting of exterior vinyl siding
#14	Hose line extending up interior stairwell; note accumulation of debris
#15	View up the stairs from first floor; note deep charring on stairs
#16	View from second floor down the stairs
#17	View from second floor down the stairs
#18	View from underneath the stairs
#19	Day time view of base of stairs, which has been cleaned to the floor. Fifth victim was discovered at this location by Engine 201

#20	Second day time view of base of stairs and view of southern exposure
#21	Location of victims one and two (ultimately six) just inside landing of stairs on second floor, note lack of burns and smoke pattern on carpet
#22	Additional view of victims one, two, and six; note lack of burn/smoke on baseboard due to adult female's head coming to rest against wall
#23	Additional view of the location of victims one, two, and six
#24	View of rear window on east side alleged to be broken by male survivor; victims three and four found just inside the window
#25	Additional view of rear window
#26	View of tarp covering bodies of victims three and four
#27	View of tarp covering bodies of victims three and four and relationship to rear window
#28	Point of origin on front porch
#29	Debris at point of origin
#30	Northwest corner of front porch; note debris
#31	Electrical outlet and battery charger at point of origin
#32	Porch cleaned to floor
#33	Southwest corner of porch cleaned to floor
#34	Southeast corner of porch prior to cleaning; note stairs just inside
#35	Electrical outlet with battery charge still plugged into the outlet
#36	Close-up of the outlet
#37	Close-up of the outlet, with plug still in tack
#38	See number #37
#39	Plug removed
#40	Point of origin, with outlet box
#41	Outlet
#42	Teddy bear memorial left at scene
#43	Teddy bear memorial left at scene

The following photos were taken after the incident by John Lee Cook, Jr.:

#44	North exposure: note melted siding
#45	Front of house: note sign advertising the relief fund
#46	Rear of house: note appliances on back porch
#47	South side of residence

#1　Night view of northwest corner (front) of house

#2　Night view of front entrance on west side of house; note firefighters and hoselines on front porch

#3   Daylight view of front of house; investigators at work

#4   View of north side of house; note multiple roof lines indicating possible enlargement of second floor

#5 Day time view of south side of structure; open back porch is located behind lattice; note electric meter in center of wall to the left of the second window

#6 Day time view of rear of house and open back porch; note: this is where male occupant climbed up and broke out rear window and called for children; two bodies located just inside the rear window on second floor

#7   Exterior view of enclosed front porch

#8   Exterior view of second floor window of rear bedroom; note charring pattern

#9   North view of exterior; note ground ladder extended to second floor

#10   Another view of north wall of house; note ladder and char patterns above all three windows

#11 Night view of southern wall; again pattern of exterior side indicative of former enlargement of second floor

#12   Northern exposure; note melting of exterior vinyl siding

#13   Northern exposure; note melting of exterior vinyl siding

#14 Hose line extending up interior stairwell; note accumulation of debris

#15   View up the stairs from first floor; note deep charring on stairs

#16　View from second floor down the stairs

#17 View from second floor down the stairs

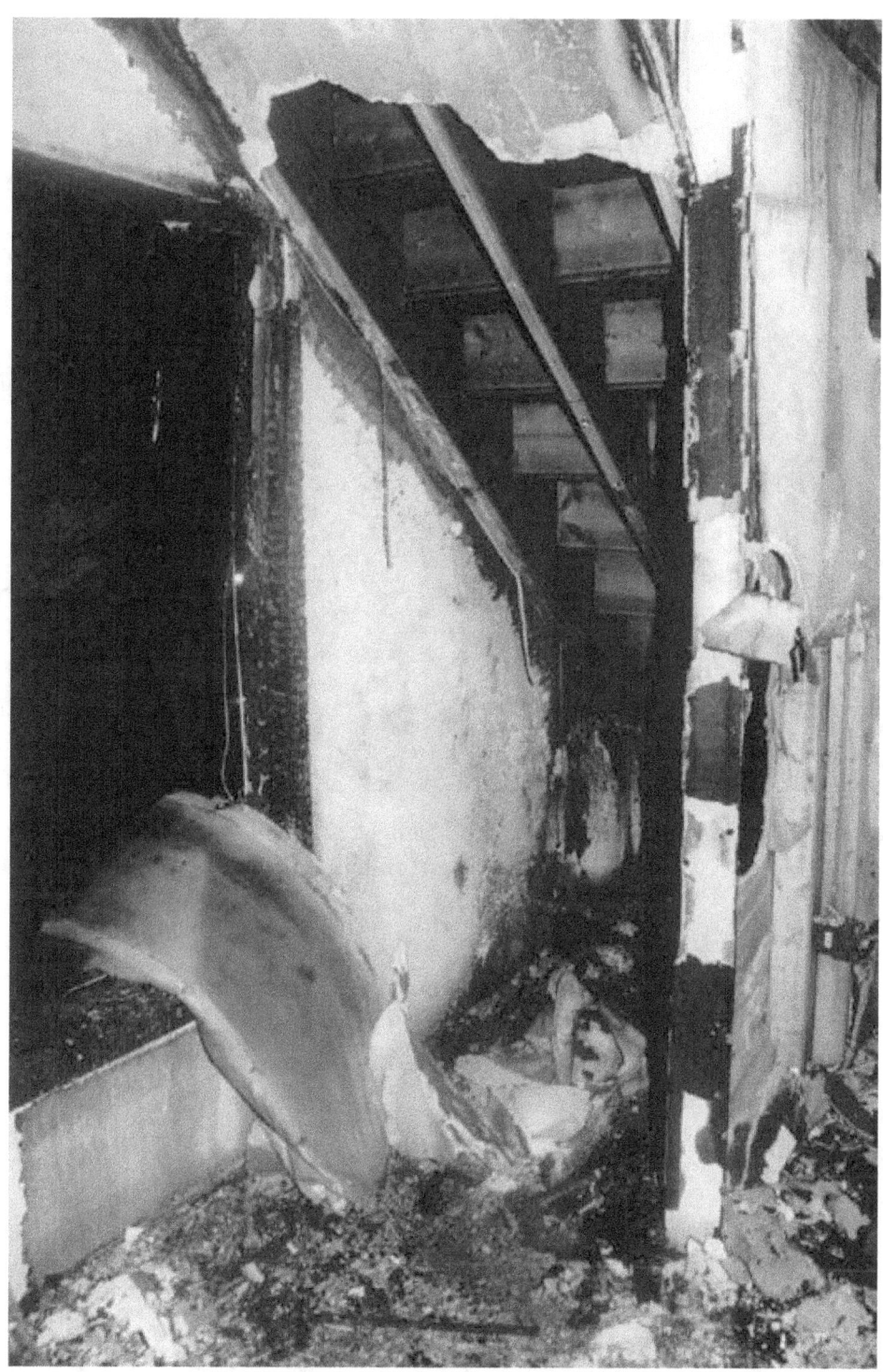

#18 View from underneath the stairs

#19 Day time view of base of stairs, which has been cleaned to the floor. Fifth victim was discovered at this location by Engine 201

#20 Second day time view of base of stairs and view of southern exposure

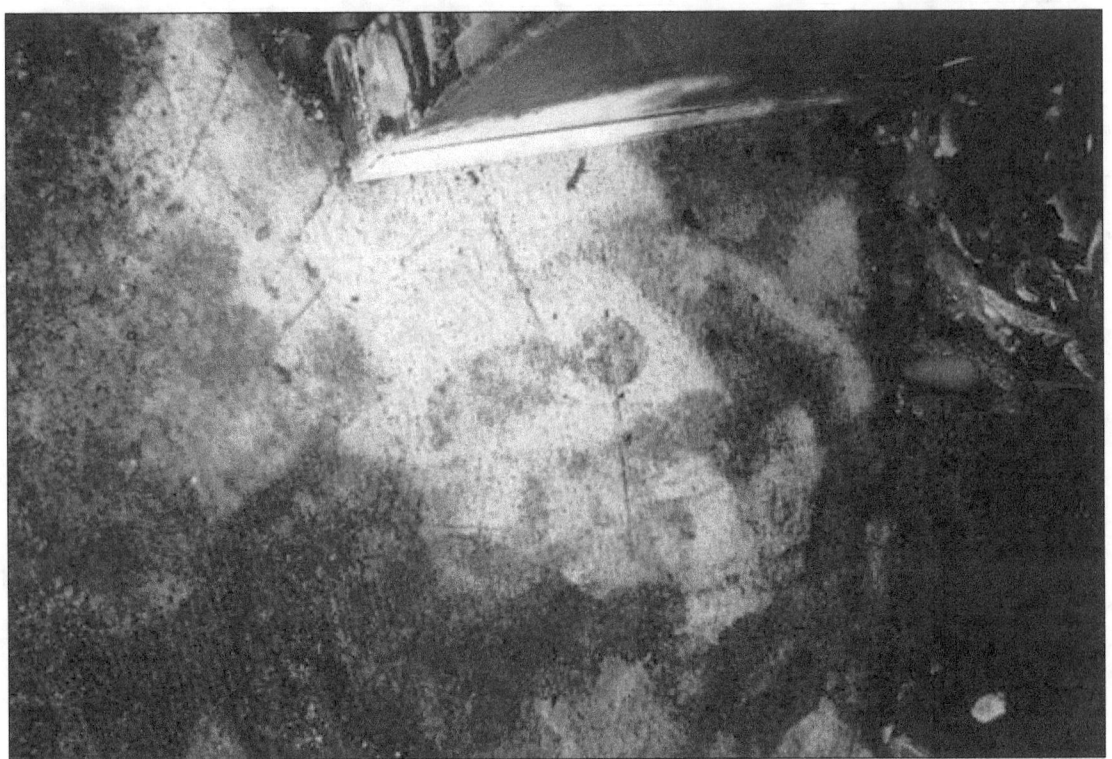

#21 Location of victims one and two (ultimately six) just inside landing of stairs on second floor, note lack of burns and smoke pattern on carpet

#22 Additional view of victims one, two, and six; note lack of burn/smoke on baseboard due to adult female's head coming to rest against wall

#23 Additional view of the location of victims one, two, and six

#24 View of rear window on east side alleged to be broken by male survivor; victims three and four found just inside the window

#25   Additional view of rear window

#26 View of tarp covering bodies of victims three and four

#27 View of tarp covering bodies of victims three and four and relationship to rear window

#28 Point of origin on front porch

#29   Debris at point of origin, window on left

#30 Northwest corner of front porch; note debris

#31   Electrical outlet and battery charger at point of origin

#32   Porch cleaned to floor

#33 Southwest corner of porch cleaned to floor

#34 Southeast corner of porch prior to cleaning; note stairs just inside

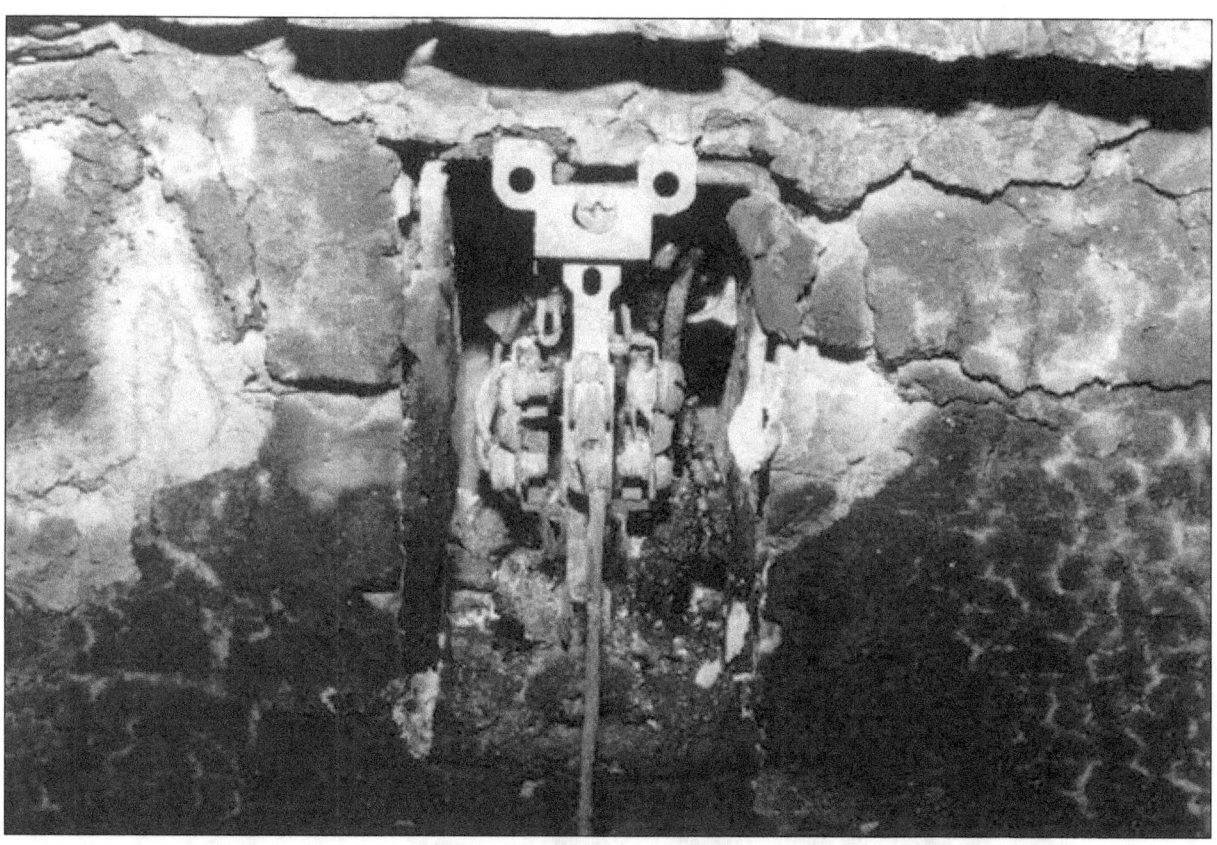

#35 Electrical outlet with battery charge still plugged into the outlet

#36   Close-up of the outlet

#37 Close-up of the outlet, with plug still in tack

#38 See number #37

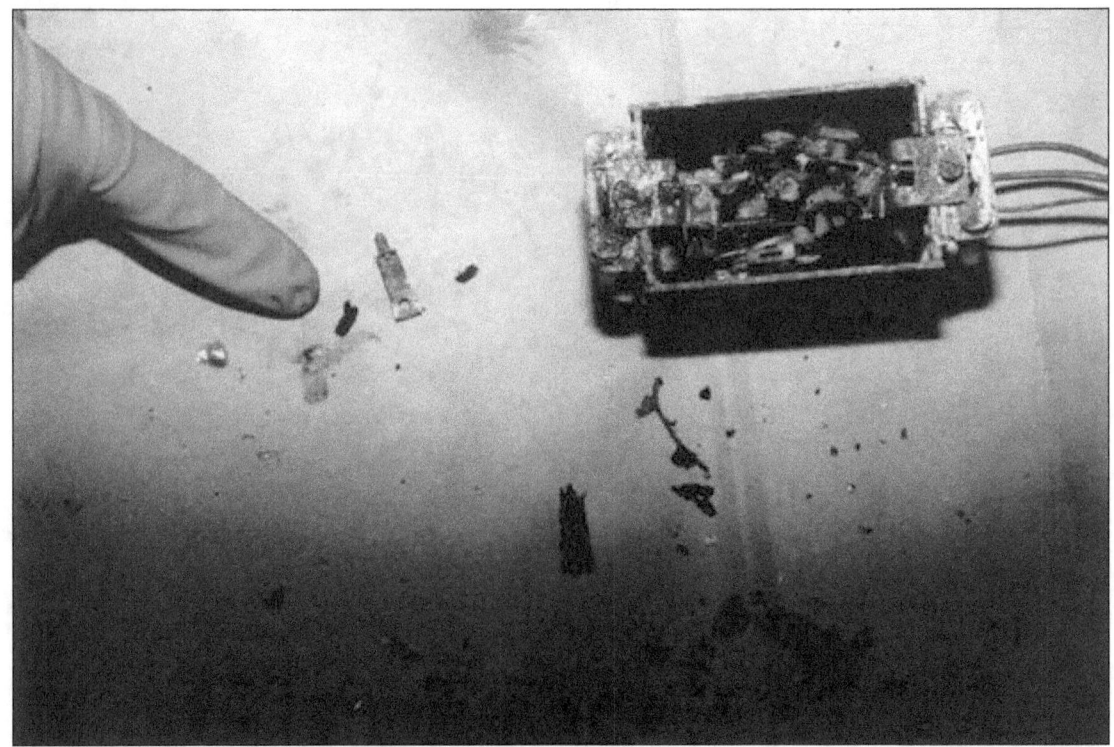

#39   Plug removed

#40   Point of origin, with outlet box

#41   Outlet

#42   Teddy bear memorial left at scene

#43 Teddy bear memorial left at scene

#44   North exposure: note melted siding

#45   Front of house: note sign advertising the relief fund

#46 Rear of house: note appliances on back porch

#47 South side of residence